A Process Algebra

For

Systems Architecture

-- The Structure-Behavior Coalescence Approach --

William S. Chao

Structure-Behavior Coalescence

Systems Architecture = **Systems Structure** + **Systems Behavior**

CONTENTS

CONTENTS...5

ABOUT THE AUTHOR...9

PART I: WHAT IS PROCESS ALGEBRA? 11

Algebraic Approach to the Study of Concurrent Systems *13*

Examples of Process Algebras *14*

Infinite-Queue SBC Process Algebra..*15*

PART II: MATHEMATICS OF INFINITE-QUEUE SBC

PROCESSES 17

Interaction *19*

Sequentialization of Interactions..*22*

Parallel Composition of Processes *23*

Process Replication..*24*

Conditional Definition of a Process...*25*

Null Process *26*

PART III: THE STRUCTURE-BEHAVIOR COALESCENCE

APPROACH 27

Structure-Behavior Coalescence Means to Integrate the Systems

Structure and Systems Behavior...*29*

Core Theme of Structure-Behavior Coalescence *30*

Interactions among Components and Actors to Draw Forth the

Systems Behavior..*32*

Collection of All Interaction Flow Diagrams Defines the Systems

Architecture *33*

PART IV: LANGUAGE CONSTRUCTS OF INFINITE-

QUEUE SBC PROCESS ALGEBRA..**37**

Backus-Naur Form of Infinite-Queue SBC Processes *39*

Parallel of All Replicated Interaction Flow Diagrams Defines the Infinite-Queue SBC Process of Systems Architecture *41*

A NULL Process or the Parallel Composition Composing All Replicated Interaction Flow Diagrams Defines the Parallel of All Replicated Interaction Flow Diagrams *42*

Parallel Composition of a Countably Infinite Number of an Interaction Flow Diagram Defines the Replicated Interaction Flow Diagram *43*

An Interaction Flow Diagram Consists of a Type_1 Expression, Followed by a Sequential Composition, and Followed by Zero or More Expressions *44*

Zero or More Expressions either are a Null Process or Consist of a Type_1_Or_2 Expression, Followed by a Sequential Composition, and Followed by Zero or More Expressions *45*

Type_1_Or_2 Expression is either Type_1 or Type_2 ....*46*

Type_1 Expression is either a Type_1 Interaction or a Conditional Type_1 Interaction *47*

Type_2 Expression is either a Type_2 Interaction or a Conditional Type_2 Interaction *48*

An Actor Interacting with a Component Defines the Type_1 Interaction *49*

A Component Interacting with another Component Defines the Type_2 Interaction *50*

PART V: TRANSITIONAL SEMANTICS OF INFINITE-QUEUE SBC PROCESS ALGEBRA..**51**

Transitional Semantics...*53*

Rule of Prefix..*56*

Rules of Parallel Composition *57*

Rule of Constants *58*

Examples of Transitional Semantics *59*

PART VI: FIRST EXAMPLE -- INFINITE-QUEUE SBC PROCESS OF THE ROBOT SYSTEM **63**

Systems Architecture of the Robot System *65*

BNF Tree of the Robot System *68*

Interactions of the Robot System *69*

Interaction Flow Diagrams of the Robot System *71*

Replicated Interaction Flow Diagrams of the Robot System *72*

Infinite-Queue SBC Process of the Robot System *74*

PART VII: SECOND EXAMPLE -- INFINITE-QUEUE SBC PROCESS OF THE AUTOMOBILE SYSTEM **77**

Systems Architecture of the Automobile System *79*

BNF Tree of the Automobile System *82*

Interactions of the Automobile System *84*

Interaction Flow Diagrams of the Automobile System *86*

Replicated Interaction Flow Diagrams of the Automobile System . *87*

Infinite-Queue SBC Process of the Automobile System *89*

ABOUT THE AUTHOR

Dr. William S. Chao is the CEO & founder of SBC Architecture International®. SBC (Structure-Behavior Coalescence) architecture is a systems architecture which demands the integration of systems structure and systems behavior of a system. SBC architecture applies to hardware architecture, software architecture, enterprise architecture, knowledge architecture and thinking architecture. The core theme of SBC architecture is: Architecture = Structure + Behavior.

William S. Chao received his bachelor degree (1976) in telecommunication engineering and master degree (1981) in information engineering, both from the National Chiao-Tung University, Taiwan. From 1976 till 1983, he worked as an engineer at Chung-Hwa Telecommunication Company, Taiwan.

William S. Chao received his master degree (1985) in information science and Ph.D. degree (1988) in information science, both from the University of Alabama at Birmingham, USA. From 1988 till 1991, he worked as a computer scientist at GE Research and Development Center, Schenectady, New York, USA.

Dr. William S. Chao has been teaching at National Sun Yat-

Sen University, Taiwan since 1992 and now serves as the president of Association of Enterprise Architects, Taiwan Chapter. His research covers: systems architecture, hardware architecture, software architecture, enterprise architecture, knowledge architecture and thinking architecture.

PART I: WHAT IS PROCESS ALGEBRA?

Algebraic Approach to the Study of Concurrent Systems

Process algebras are a diverse family of related approaches to the study of concurrent systems. Their tools are algebraic languages for the high-level description of interactions, communications and synchronizations between a collection of independent agents or processes.

Process algebras also provide algebraic laws that allow process descriptions to be manipulated and analyzed, and permit formal reasoning about equivalences and observation congruence among processes.

Examples of Process Algebras

There are several leading algebraic approaches to modeling concurrent systems.

Communicating Sequential Processes (CSP) was first described in a 1978 paper by C. A. R. Hoare.

Arthur John Robin Gorell Milner introduced the Calculus of Communicating Systems (CCS) around 1980.

Algebra of Communicating Processes (ACP) was initially developed by Jan Bergstra and Jan Willem Klop in 1982.

Infinite-Queue SBC Process Algebra

Single-queue SBC process algebra, multi-queue SBC process algebra and infinite-queue SBC process algebra are the three SBC process algebras.

Infinite-queue SBC process algebra evolved from CCS (Calculus of Communicating Systems).

CCS is a general process algebra language for the study of concurrent systems. Unlike CCS, infinite-queue SBC process algebra is only applicable to systems architecture.

PART II: MATHEMATICS OF INFINITE-QUEUE SBC PROCESSES

Interaction

An interaction represents an indivisible and instantaneous handshake or communication between two agents. The caller agent (either external environment's actor or component) communicates with the callee agent (component) through the interaction.

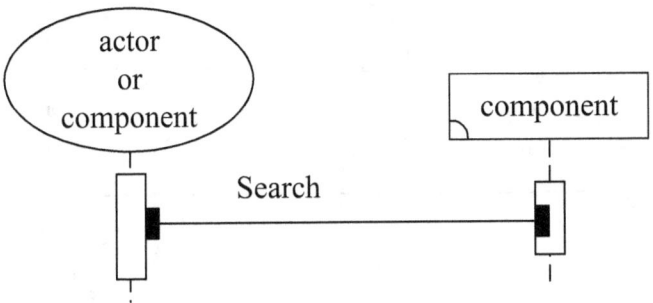

There are two ports, i.e., calling port or called port, of an interaction. The caller agent owns the "calling port" of the interaction.

The caller agent together with the "calling port" is named the "calling action".

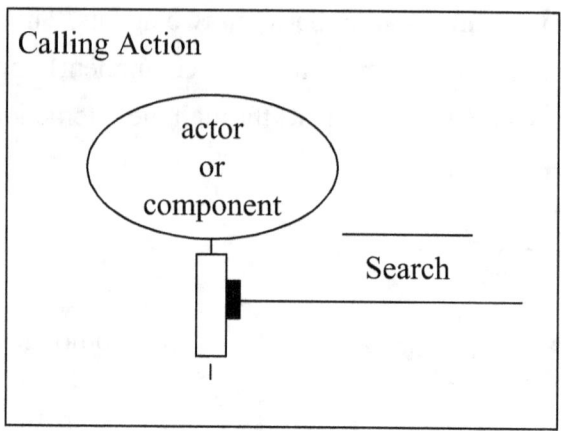

The callee agent owns the "called port" of the interaction.

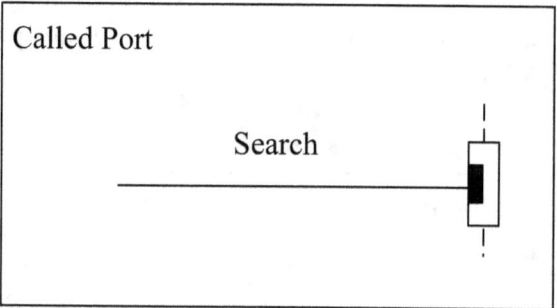

The callee agent together with the "called port" is named the "called action".

In order to simplify the interaction diagram, we will redraw it as follows.

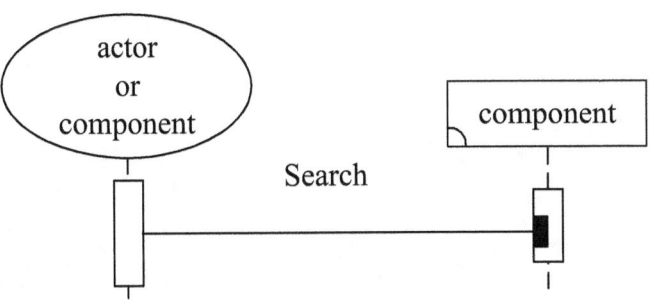

Sequentialization of Interactions

Sometimes interactions must be temporally ordered. For example, it might be desirable to specify algorithms such as: execute the "a" interaction first and then execute the "P" process later. Sequentialization of interactions can be used for such purposes.

Sequentialization of interactions, usually written as the a$\bullet P$ process, indicates that it will perform the "a" interaction first and continue as the "P" process.

Parallel Composition of Processes

Parallel composition of two processes P and Q, usually written $P\|Q$, is the key primitive distinguishing the process algebras from sequential models of process executions.

Parallel composition allows the executions in P and Q to proceed simultaneously and independently.

Process Replication

The operators presented so far describe only finite interaction and are consequently insufficient for full computability, which includes non-terminating behavior. Replication is the operator that allows finite descriptions of infinite behavior.

For example, replication $!P$ can be understood as abbreviating the parallel composition of a countably infinite number of P processes.

Conditional Definition of a Process

A process can be defined by a one-or-more-armed conditional expression.

For example, the process (**if** $cond_1$ **then** P_1)+(**if** $cond_2$ **then** P_2)...+(**if** $cond_j$ **then** P_j) will proceed as the process P_1 if the "$cond_1$" value is true, or proceed as the process P_2 if the "$cond_2$" value is true,..., or proceed as the process P_j if the "$cond_j$" value is true.

Null Process

Process algebras generally also include a null process, denoted as *STOP*, which has no interaction points. It is utterly inactive and its sole purpose is to act as the inductive anchor on top of which more interesting processes can be generated.

The process "*STOP•P*" (i.e. sequential composition of processes *STOP* and *P*) equals to the process "*STOP*".

$$STOP \bullet P \quad = \quad STOP$$

The process "*P‖STOP*" (i.e. parallel composition of processes *P* and *STOP*) equals to the process "*STOP‖P*" (i.e. parallel composition of processes *STOP* and *P*) which equals to the process "*P*".

$$P \parallel STOP \quad = \quad STOP \parallel P \quad = \quad P$$

PART III: THE STRUCTURE-BEHAVIOR COALESCENCE APPROACH

Structure-Behavior Coalescence Means to Integrate the Systems Structure and Systems Behavior

Systems structure and systems behavior are the two most prominent views of a system, integrating the systems structure and systems behavior is apparently the best way to achieve an integrated whole of a system.

If we are not able to integrate the systems structure and systems behavior, then there is no way that we are able to integrate the whole system.

Structure-behavior coalescence (SBC) provides an elegant way to integrate the systems structure and systems behavior of a system. In other words, SBC facilitates an integrated whole of a system.

Core Theme of Structure-Behavior Coalescence

The core theme of structure-behavior coalescence is: "Systems Architecture = Systems Structure + Systems Behavior."

Systems Structure X ┼ Systems Behavior X

One systems structure will draw forth one systems behavior. That is, the systems behavior is attached to or built on the systems structure in the SBC approach.

In other words, the systems behavior can not exist alone; it must be loaded on the systems structure just like a cargo is loaded on a ship.

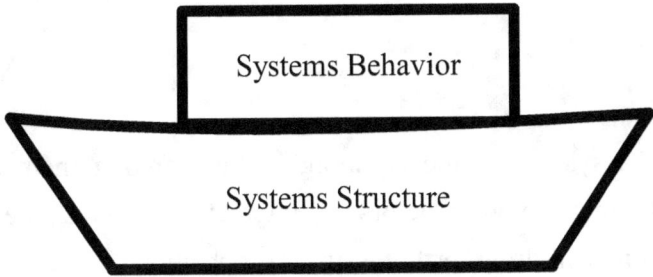

Interactions among Components and Actors to Draw Forth the Systems Behavior

In a system, if the components, and among them and the external environment's actors to interact (or handshake), these interactions will draw forth the systems behavior.

We conclude that "interaction" plays an important factor in integrating the systems structure and systems behavior for a system.

The overall behavior of a system consists of many individual behaviors. Each individual behavior represents an execution path. We use an interaction flow diagram (IFD) to demonstrate this individual behavior.

Collection of All Interaction Flow Diagrams Defines the Systems Architecture

The collection of all interaction flow diagrams defines the integration of systems structure and systems behavior of a system.

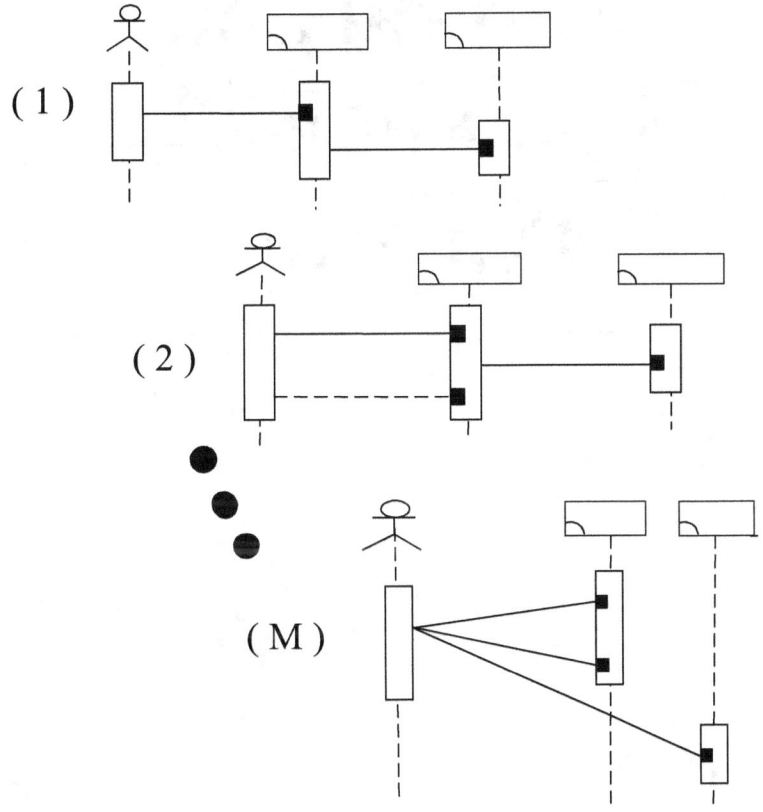

That is, the collection of all interaction flow diagrams defines the systems architecture.

Systems architecture (SA) represents a knowledge repository of a system. Stakeholders can submit and acquire knowledge to and from this knowledge repository.

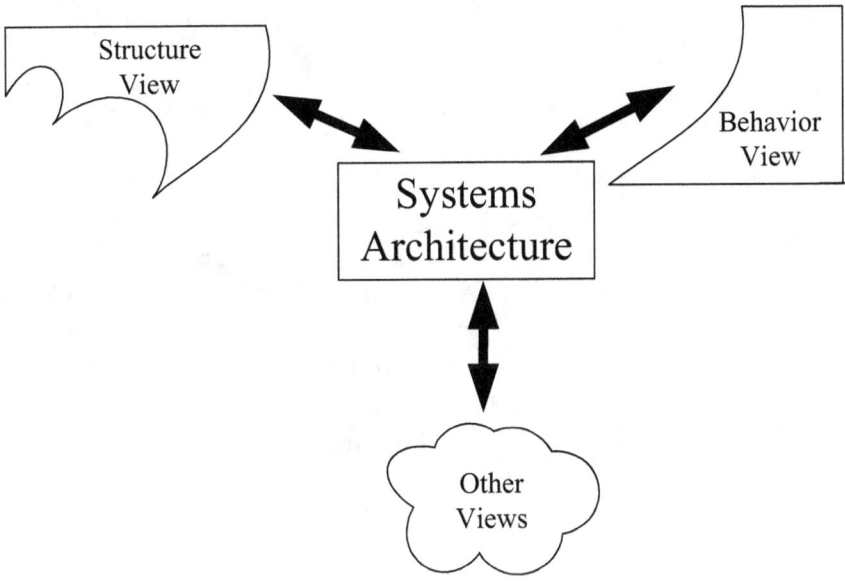

So, the collection of all interaction flow diagrams represents a knowledge repository of a system. Stakeholders can submit and acquire knowledge to and from this knowledge repository.

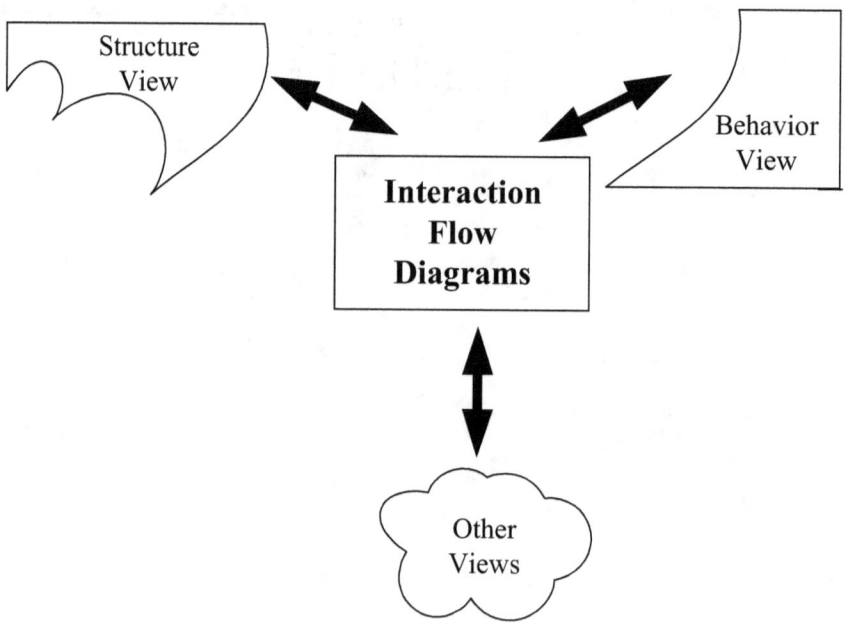

PART IV: LANGUAGE CONSTRUCTS OF INFINITE-QUEUE SBC PROCESS ALGEBRA

Backus-Naur Form of Infinite-Queue SBC Processes

The set of infinite-queue SBC (i.e. Structure-Behavior Coalescence) processes for systems architecture is defined by the following BNF grammar:

(1) <Infinite-Queue_SBC_Process_of_Systems_Architecture> ::=
    <Parallel_of_!IFD>

(2) <Parallel_of_!IFD> ::=   *STOP*
  | <!IFD> "‖" <Parallel_of_!IFD>

(3) <!IFD> ::=   <IFD>
  | <IFD> "‖" <!IFD>

(4) <IFD> ::=
    <Type_1_Expression> "●" <Zero_Or_More_Expressions>

(5) <Zero_Or_More_Expressions> ::=   *STOP*
  | <Type_1_Or_2_Expression> "●" <Zero_Or_More_Expressions>

(6) <Type_1_Or_2_Expression>   ::= <Type_1_Expression>
  | <Type_2_Expression>

(7) <Type_1_Expression>   ::= <Type_1_Interaction>
  | <Condition> <Type_1_Interaction>
    {"+" <Condition> <Type_1_Interaction>}

(8) <Type_2_Expression>   ::= <Type_2_Interaction>
  | <Condition> <Type_2_Interaction>
    {"+" <Condition> <Type_2_Interaction>}

(9) <Type_1_Interaction> ::= <Actor> <Operation_Call_Or_Return>
    <Operation_Call_Or_Return_Formula> <Component>

(10) <Type_2_Interaction> ::= <Component> <Operation_Call_Or_Return>
    <Operation_Call_Or_Return_Formula> <Component>

Parallel of All Replicated Interaction Flow Diagrams Defines the Infinite-Queue SBC Process of Systems Architecture

Rule 1 describes that the parallel of all replicated interaction flow diagrams (i.e. Parallel_of_!IFD), in which each interaction flow diagram may replicate itself a countably infinite times (i.e. !IFD), defines the infinite-queue SBC process of systems architecture.

Rule 1
<Infinite-Queue_SBC_Process_for_Systems_Architecture> ::=   <Parallel_of_!IFD>

A NULL Process or the Parallel Composition Composing All Replicated Interaction Flow Diagrams Defines the Parallel of All Replicated Interaction Flow Diagrams

Rule 2 describes that we use either a) a null process (i.e. *STOP*), or b) the parallel composition (i.e. ||) composing all replicated interaction flow diagrams (i.e. !IFD), to define the parallel of all replicated interaction flow diagrams (i.e. Parallel_of_!IFD).

Rule 2

<Parallel_of_!IFD> ::=
     STOP
   | <!IFD> "||" <Parallel_of_!IFD>

Parallel Composition of a Countably Infinite Number of an Interaction Flow Diagram Defines the Replicated Interaction Flow Diagram

Rule 3 describes that the parallel composition (i.e. ∥) of a countably infinite number of an interaction flow diagram (i.e. IFD) defines the replicated interaction flow diagram (i.e. !IFD).

Rule 3
<!IFD> ::=   <IFD> | <IFD> "∥" <!IFD>

An Interaction Flow Diagram Consists of a Type_1 Expression, Followed by a Sequential Composition, and Followed by Zero or More Expressions

Rule 4 describes that an interaction flow diagram (i.e. IFD) consists of a type_1 expression (i.e. Type_1_Expression), followed by a sequential composition (i.e. ●), and followed by zero or more expressions (i.e. Zero_Or_More_Expressions).

Rule 4
<IFD> ::=   <Type_1_Expression> "●" <Zero_Or_More_Expressions>

Zero or More Expressions either are a Null Process or Consist of a Type_1_Or_2 Expression, Followed by a Sequential Composition, and Followed by Zero or More Expressions

Rule 5 describes that zero or more expressions (i.e. Zero_Or_More_Expressions) either a) are a null process (i.e. *STOP)*, or b) consist of a type_1_or_2 expression (i.e. Type_1_Or_2_Expression), followed by a sequential composition (i.e. •), and followed by zero or more expressions (i.e. Zero_Or_More_Expressions).

Rule 5
<Zero_Or_More_Expressions> ::=
STOP
| <Type_1_Or_2_Expression> "•" <Zero_Or_More_Expressions>

Type_1_Or_2 Expression is either Type_1 or Type_2

Rule 6 describes that the type_1_or_2 expression (i.e. Type_1_Or_2_Expression) is either a type_1 expression (i.e. Type_1_Expression) or a type_2 expression (i.e. Type_2_Expression).

Rule 6
<Type_1_Or_2_Expression> ::=   <Type_1_Expression> | <Type_2_Expression>

Type_1 Expression is either a Type_1 Interaction or a Conditional Type_1 Interaction

Rule 7 describes that the type_1 expression (i.e. Type_1_Expression) is either a type_1 interaction (i.e. Type_1_Interaction) or a conditional type_1 interaction (i.e. one-or-more-armed conditional expression of Type_1_Interaction).

Rule 7
<Type_1_Expression> ::= <Type_1_Interaction> | <Condition> <Type_1_Interaction> 　　{"+" <Condition> <Type_1_Interaction>}

Type_2 Expression is either a Type_2 Interaction or a Conditional Type_2 Interaction

Rule 8 describes that the type_2 expression (i.e. Type_2_Expression) is either a type_2 interaction (i.e. Type_2_Interaction) or a conditional type_2 interaction (i.e. one-or-more-armed conditional expression of Type_2_Interaction).

Rule 8

<Type_2_Expression>   ::= <Type_2_Interaction>
  | <Condition> <Type_2_Interaction>
     {"+" <Condition> <Type_2_Interaction>}

An Actor Interacting with a Component Defines the Type_1 Interaction

Rule 9 describes that an actor interacting with a component defines the type_1 interaction.

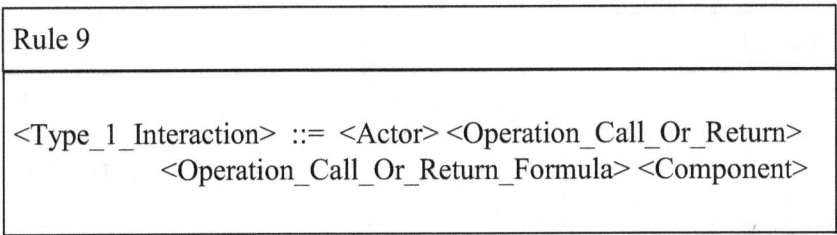

Rule 9
<Type_1_Interaction> ::= <Actor> <Operation_Call_Or_Return> <Operation_Call_Or_Return_Formula> <Component>

A Component Interacting with another Component Defines the Type_2 Interaction

Rule 10 describes that a component interacting with another component defines the type_2 interaction.

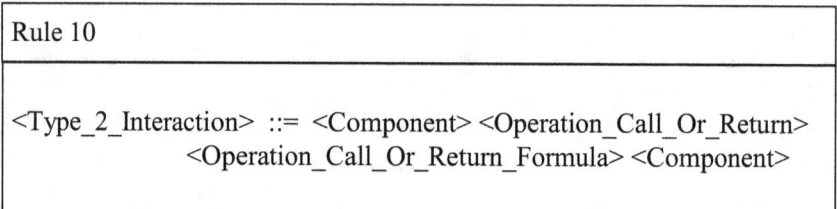

Rule 10

<Type_2_Interaction> ::= <Component> <Operation_Call_Or_Return> <Operation_Call_Or_Return_Formula> <Component>

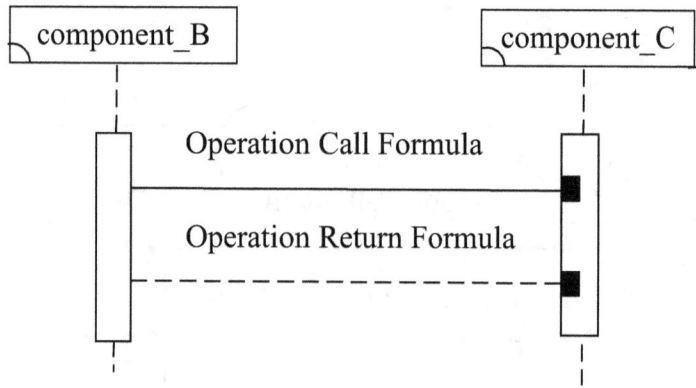

PART V: TRANSITIONAL SEMANTICS OF INFINITE-QUEUE SBC PROCESS ALGEBRA

Transitional Semantics

We assume an infinite set Δ of type_1_or_2 interactions, and use a_1, a_2...to range over Δ. Further, we let Φ be the set of process Constants, and use A_1, A_2...to range over Φ. We let Π be the set of processes, and use P_1, Q_1...to range over Π. We let Ψ be the set of process expressions, and use E_1, E_2...to range over Ψ.

Entity set	Entity name	Type of entity
Δ	$a_1, a_2...$	type_1_or_2 interactions
Φ	$A_1, A_2...$	process Constants
Π	$P_1, Q_1...$	processes
Ψ	$E_1, E_2...$	process expressions

In giving meaning to the infinite-queue SBC process algebra, we shall use the following labelled transition system (LTS)

$$(\Psi, \Delta, \rightarrow)$$

which consists of a set Ψ of process expressions, a set Δ of "type 1 or 2 interactions", and a transition relation $\rightarrow \subseteq \Psi X \Delta X \Psi$ where

$(E_i, a, E_j) \in \rightarrow$ is denoted by $E_i \overset{a}{\rightarrow} E_j$.

The semantics for Ψ consists in the transition rules of each transition relation $\rightarrow$ over $\Psi X \Delta X \Psi$. These transition rules will follow the structure of expressions.

We give the complete set of transition rules; the names Prefix, Parallel, and Constant indicate that the rules are associated respectively with Prefix, Parallel Composition and with Constants.

Prefix	$$\dfrac{}{a \bullet E \xrightarrow{a} E}$$
Parallel$_1$	$$\dfrac{E \xrightarrow{a} E'}{E \| F \xrightarrow{a} E' \| F}$$
Parallel$_2$	$$\dfrac{F \xrightarrow{a} F'}{E \| F \xrightarrow{a} E \| F'}$$
Constant	$$\dfrac{P \xrightarrow{a} P'}{A \xrightarrow{a} P'} \quad (A \overset{\text{def}}{=} P)$$

Rule of Prefix

The rule for Prefix can be read as follows: Under any circumstances, we always infer $a \bullet E \xrightarrow{a} E$. That is, an expression, with an interaction prefixed to it, will use this interaction to accomplish the transition.

$$\frac{}{a \bullet E \xrightarrow{a} E}$$

Rules of Parallel Composition

There are two transition rules for parallel composition. Rule Parallel$_1$ indicates that from $E \xrightarrow{a} E'$ we shall infer $E\|F \xrightarrow{a} E'\|F$.

$$\frac{E \xrightarrow{a} E'}{E \| F \xrightarrow{a} E' \| F}$$

Rule Parallel$_2$ indicates that from $F \xrightarrow{a} F'$ we shall infer $E\|F \xrightarrow{a} E\|F'$.

$$\frac{F \xrightarrow{a} F'}{E \| F \xrightarrow{a} E \| F'}$$

Rule of Constants

The rule for Constants can be read as follows: the rule of Constants asserts that each Constant has the same transitions as its defining expression.

$$\frac{P \xrightarrow{a} P'}{A \xrightarrow{a} P'} \quad (A \stackrel{def}{=} P)$$

Examples of Transitional Semantics

As a first example, consider the infinite-queue SBC process Constant A_1 is defined as $(b_1 \bullet b_2 \bullet b_3 \bullet STOP)$. The following transition graph shows the semantics of infinite-queue SBC process A_1.

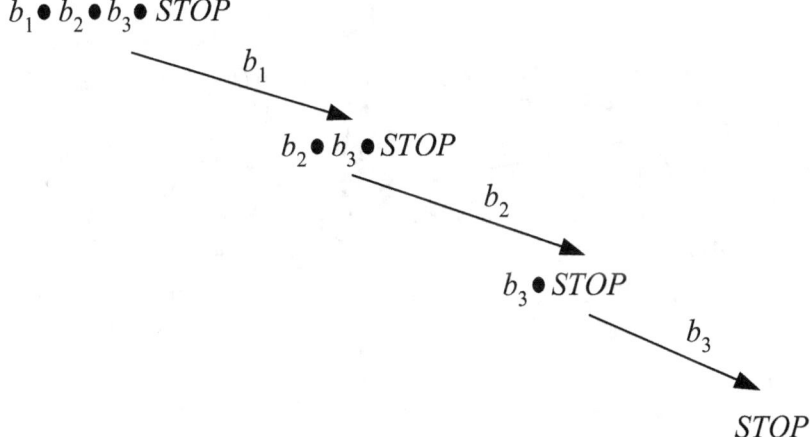

60

As a second example, consider the infinite-queue SBC process Constant A_2 is defined as $(a_1 \bullet a_2 \bullet STOP) \| (b_1 \bullet b_2 \bullet b_3 \bullet STOP)$. The following transition graph shows the semantics of infinite-queue SBC process A_2.

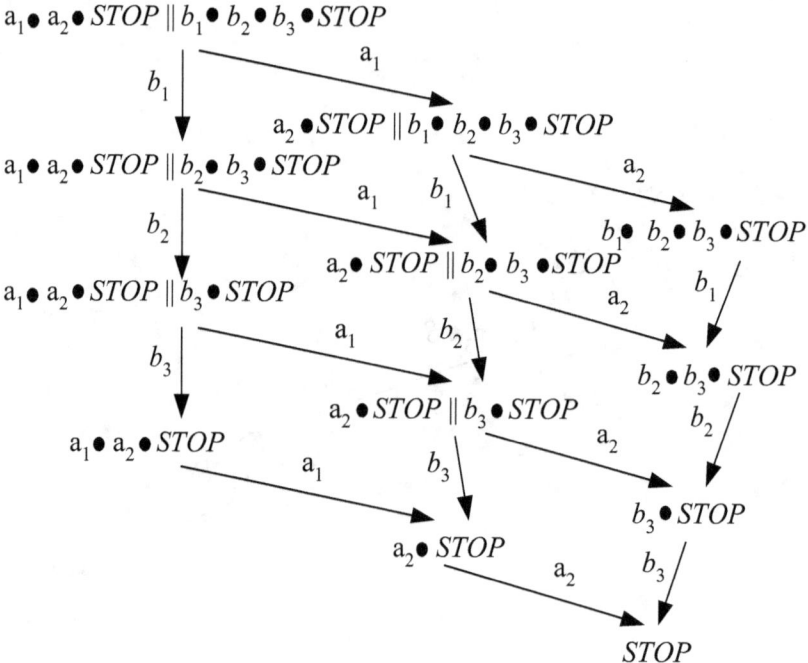

As a third example, consider the infinite-queue SBC process Constant A_3 is defined as $(b_1 \bullet STOP) \| (b_2 \bullet STOP) \| (b_3 \bullet STOP)$. The following transition graph shows the semantics of infinite-queue SBC process A_3.

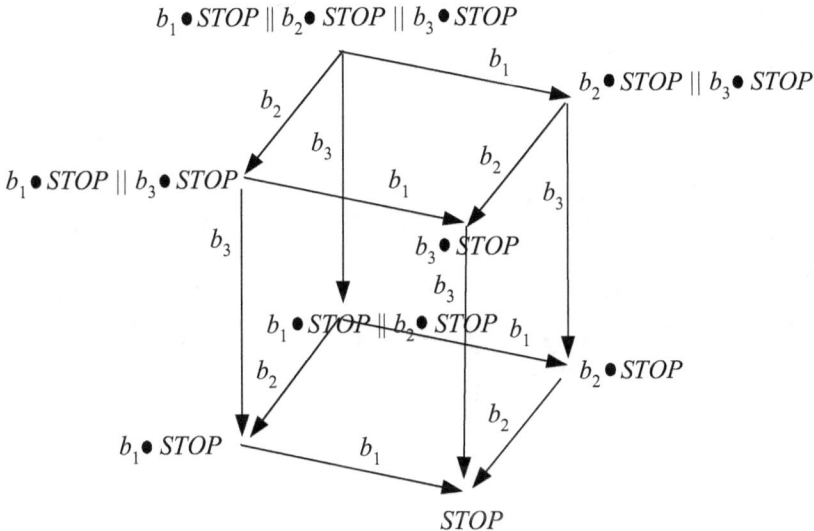

PART VI: FIRST EXAMPLE --
INFINITE-QUEUE SBC PROCESS
OF THE ROBOT SYSTEM

Systems Architecture of the Robot System

The collection of all interaction flow diagrams defines the systems architecture. The overall behavior of the *Robot* system includes two behaviors: *Writing* and *Walking*. Each of them is described by an individual IFD.

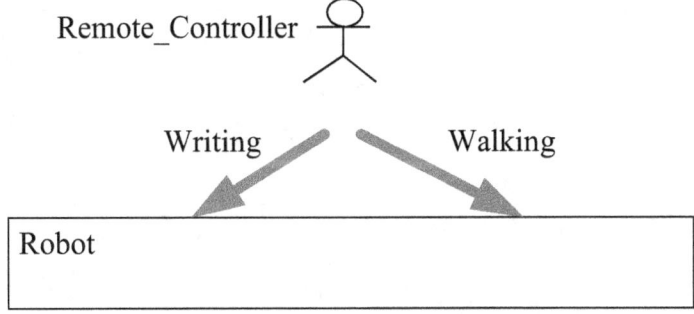

An IFD of the *Writing* behavior is shown below. First, actor *Remote_Controller* interacts with the *Head* component through the *Receive_Write_Signal* operation call interaction. Finally, component *Head* interacts with the *Hands* component through the *Move_Hand* operation call interaction.

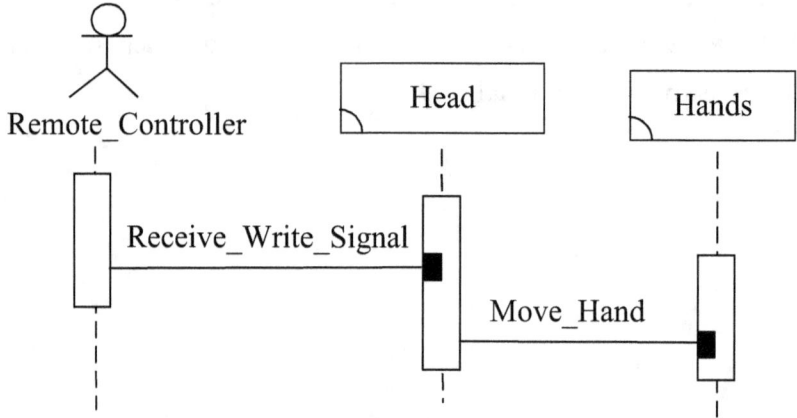

An IFD of the *Walking* behavior is shown below. First, actor *Remote_Controller* interacts with the *Head* component through the *Receive_Walk_Signal* operation call interaction. Finally, component *Head* interacts with the *Feet* component through the *Move_Foot* operation call interaction.

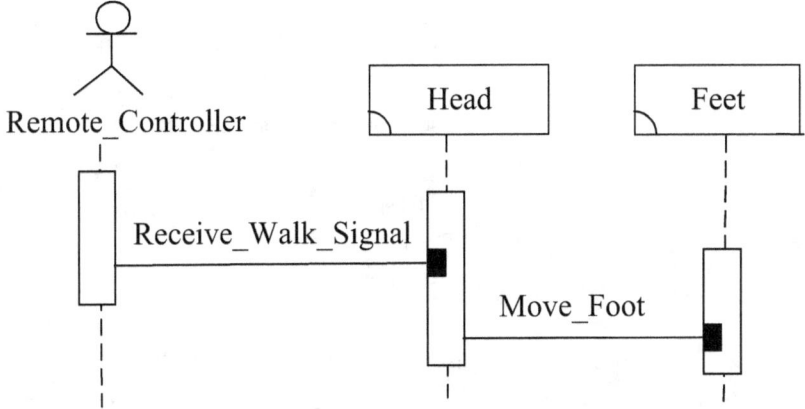

BNF Tree of the Robot System

We draw the infinite-queue SBC process algebra Backus-Naur Form tree of the *Robot* system as follows:

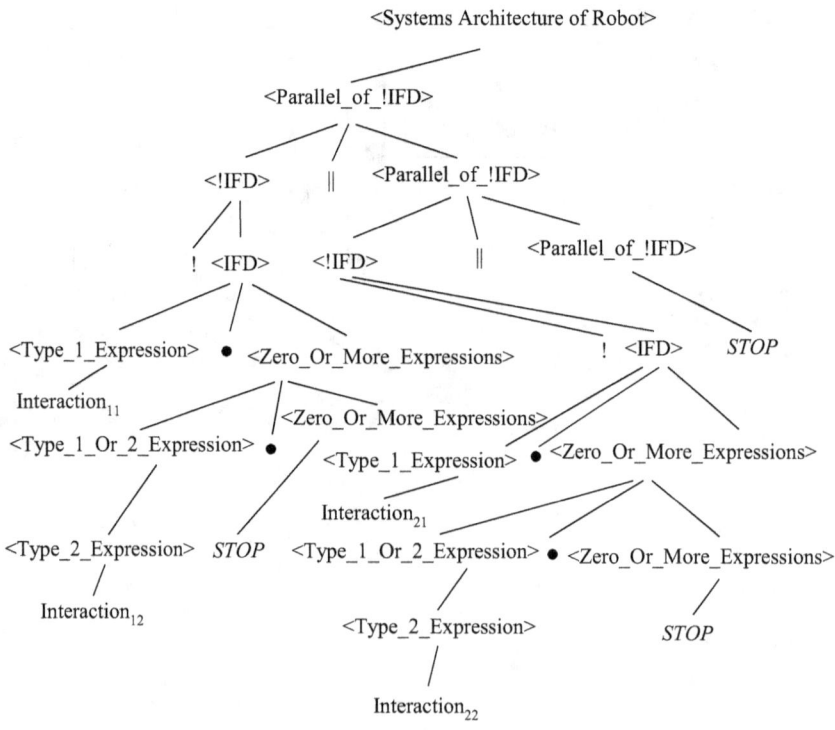

Interactions of the Robot System

Interaction$_{11}$ stands for the 1st interaction of the 1st interaction flow diagram of the *Robot* system. Interaction$_{11}$ is a type_1 interaction which describes the *Remote_Controller* actor interacts with the *Head* component.

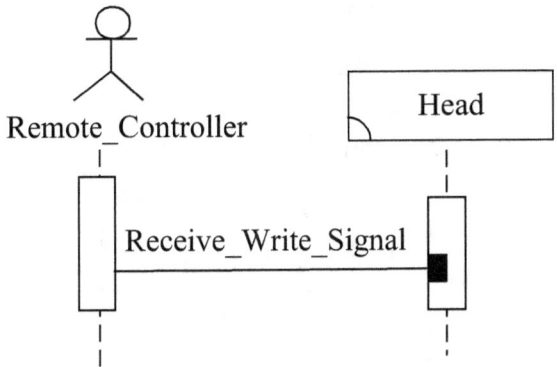

Interaction$_{12}$ stands for the 2nd interaction of the 1st interaction flow diagram of the *Robot* system. Interaction$_{12}$ is a type_2 interaction which describes the *Head* component interacts with the *Hands* component.

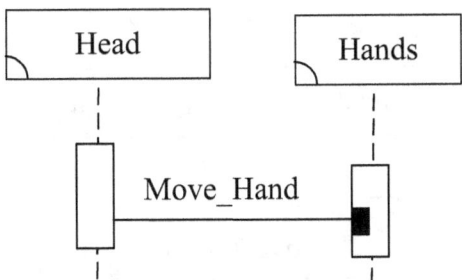

Interaction$_{21}$ stands for the 1st interaction of the 2nd interaction flow diagram of the *Robot* system. Interaction$_{21}$ is a type_1 interaction which describes the *Remote_Controller* actor interacts with the *Head* component.

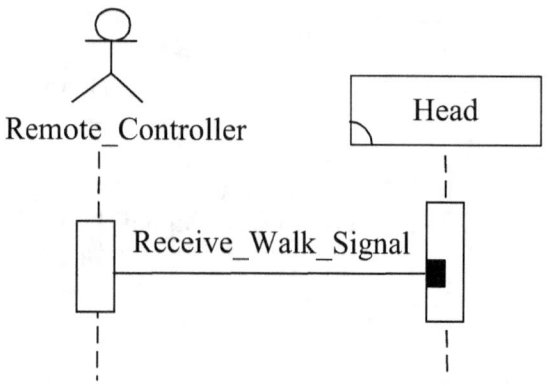

Interaction$_{22}$ stands for the 2nd interaction of the 2nd interaction flow diagram of the *Robot* system. Interaction$_{22}$ is a type_2 interaction which describes the *Head* component interacts with the *Feet* component.

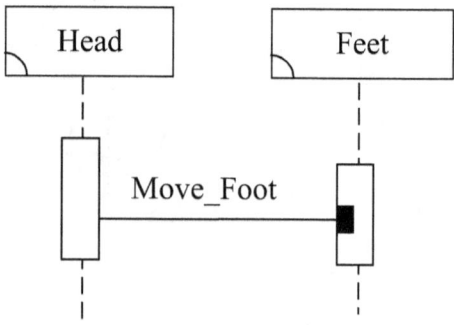

Interaction Flow Diagrams of the Robot System

IFD_1 describes the 1st interaction flow diagram, i.e. *Writing* behavior, of the *Robot* system. IFD_1 is syntactically represented as "$Interaction_{11} \bullet Interaction_{12} \bullet STOP$".

$$IFD_1 =$$
$$Interaction_{11} \bullet Interaction_{12} \bullet STOP$$

IFD_2 describes the 2nd interaction flow diagram, i.e. *Walking* behavior, of the *Robot* system. IFD_2 is syntactically represented as "$Interaction_{21} \bullet Interaction_{22} \bullet STOP$".

$$IFD_2 =$$
$$Interaction_{21} \bullet Interaction_{22} \bullet STOP$$

Replicated Interaction Flow Diagrams of the Robot System

Each interaction flow diagram may replicate itself a countably infinite time. Interaction$_{xyz}$ stands for the yth replication of the zth interaction of the xth interaction flow diagram.

!IFD$_1$ describes the infinite replication of the 1st interaction flow diagram, i.e. *Writing* behavior, of the *Robot* system. !IFD$_1$ is syntactically represented as "(Interaction$_{111}$•Interaction$_{112}$•*STOP*) || (Interaction$_{121}$•Interaction$_{122}$•*STOP*) || (Interaction$_{131}$•Interaction$_{132}$•*STOP*) .... || (Interaction$_{1\infty1}$•Interaction$_{1\infty2}$•*STOP*)".

!IFD$_1$ =

(Interaction$_{111}$ • Interaction$_{112}$ • *STOP*)
||
(Interaction$_{121}$ • Interaction$_{122}$ • *STOP*)
||
(Interaction$_{131}$ • Interaction$_{132}$ • *STOP*)
....
||
(Interaction$_{1\infty1}$ • Interaction$_{1\infty2}$ • *STOP*)

$!IFD_2$ describes the infinite replication of the 2nd interaction flow diagram, i.e. *Walking* behavior, of the *Robot* system. $!IFD_2$ is syntactically represented as "$(\text{Interaction}_{211} \bullet \text{Interaction}_{212} \bullet STOP)$ || $(\text{Interaction}_{221} \bullet \text{Interaction}_{222} \bullet STOP)$ || $(\text{Interaction}_{231} \bullet \text{Interaction}_{232} \bullet STOP)$ .... || $(\text{Interaction}_{2\infty1} \bullet \text{Interaction}_{2\infty2} \bullet STOP)$".

$$!IFD_2 =$$

$$(\text{Interaction}_{211} \bullet \text{Interaction}_{212} \bullet STOP)$$
$$||$$
$$(\text{Interaction}_{221} \bullet \text{Interaction}_{222} \bullet STOP)$$
$$||$$
$$(\text{Interaction}_{231} \bullet \text{Interaction}_{232} \bullet STOP)$$
$$.......$$
$$||$$
$$(\text{Interaction}_{2\infty1} \bullet \text{Interaction}_{2\infty2} \bullet STOP)$$

Infinite-Queue SBC Process of the Robot System

The *Robot* system's infinite-queue SBC process is syntactically represented as $(!IFD_1)\|(!IFD_2)$ which equals to "(Interaction$_{111}$●Interaction$_{112}$●*STOP*) $\|$
(Interaction$_{121}$●Interaction$_{122}$●*STOP*) $\|$
(Interaction$_{131}$●Interaction$_{132}$●*STOP*)   .... $\|$
(Interaction$_{1\infty1}$●Interaction$_{1\infty2}$●*STOP*) $\|$
(Interaction$_{211}$●Interaction$_{212}$●*STOP*) $\|$
(Interaction$_{221}$●Interaction$_{222}$●*STOP*) $\|$
(Interaction$_{231}$●Interaction$_{232}$●*STOP*)   .... $\|$
(Interaction$_{2\infty1}$●Interaction$_{2\infty2}$●*STOP*)".

Robot $\overset{\text{def}}{=\!=}$

$(\text{Interaction}_{111} \bullet \text{Interaction}_{112} \bullet STOP)$

$\parallel$

$(\text{Interaction}_{121} \bullet \text{Interaction}_{122} \bullet STOP)$

$\parallel$

$(\text{Interaction}_{131} \bullet \text{Interaction}_{132} \bullet STOP)$ ....

$\parallel$

$(\text{Interaction}_{1\infty1} \bullet \text{Interaction}_{1\infty2} \bullet STOP)$

$\parallel$

$(\text{Interaction}_{211} \bullet \text{Interaction}_{212} \bullet STOP)$

$\parallel$

$(\text{Interaction}_{221} \bullet \text{Interaction}_{222} \bullet STOP)$

$\parallel$

$(\text{Interaction}_{231} \bullet \text{Interaction}_{232} \bullet STOP)$ ....

$\parallel$

$(\text{Interaction}_{2\infty1} \bullet \text{Interaction}_{2\infty2} \bullet STOP)$

PART VII: SECOND EXAMPLE -- INFINITE-QUEUE SBC PROCESS OF THE AUTOMOBILE SYSTEM

Systems Architecture of the Automobile System

The collection of all interaction flow diagrams defines the systems architecture. The overall behavior of the *Automobile* includes two behaviors: *Accelerate_the_Car* and *Stop_the_Car*. Each of them is described by an individual IFD.

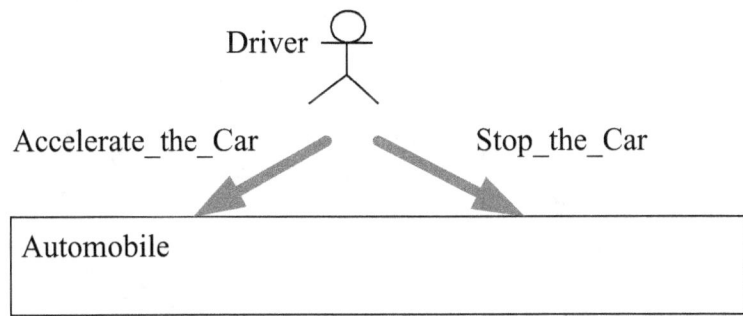

An IFD of the *Accelerate_the_Car* behavior is shown below. First, actor *Driver* interacts with the *Gas_Pedal* component through the *Depress_Gas_Pedal* operation call interaction. Finally, component *Gas_Pedal* interacts with the *Engine* component through the *Fuel_Supply* operation call interaction.

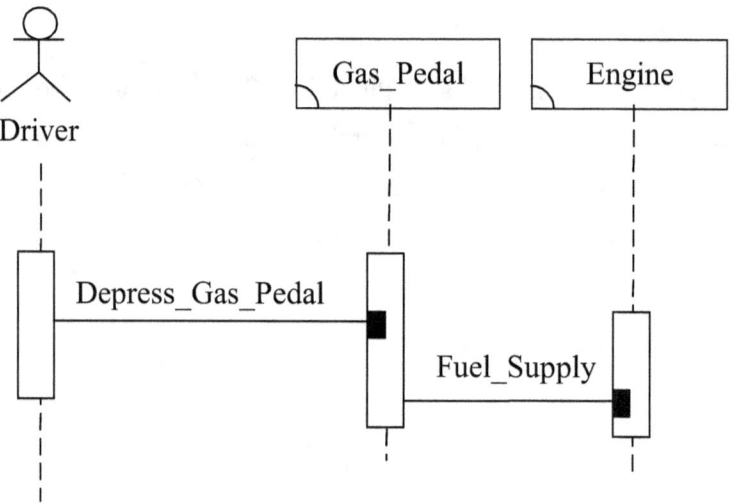

An IFD of the *Stop_the_Car* behavior is shown below. First, actor *Driver* interacts with the *Brake_Pedal* component through the *Depress_Brake_Pedal* operation call interaction. Finally, component *Brake_Pedal* interacts with the *Brake_Pad* component through the *Move_Inward* operation call interaction.

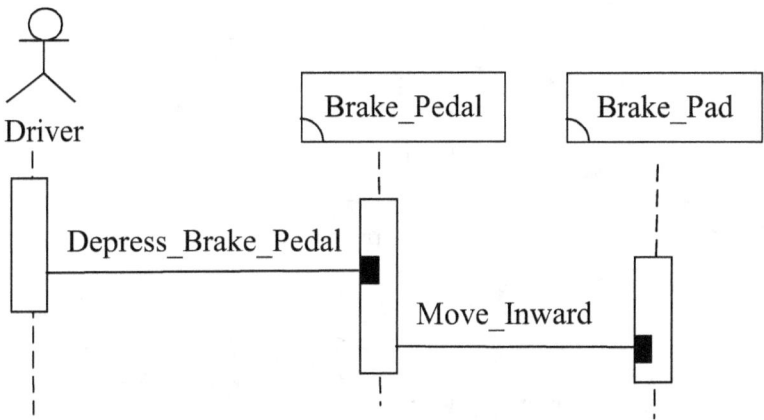

BNF Tree of the Automobile System

We draw the infinite-queue SBC process algebra Backus-Naur Form tree of the *Automobile* system as follows:

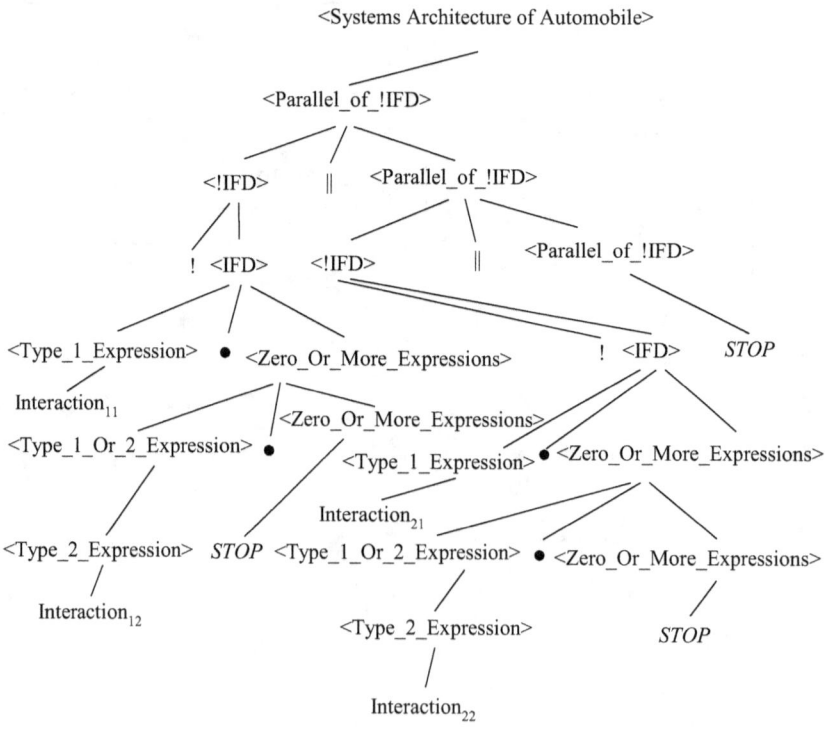

Interactions of the Automobile System

Interaction$_{11}$ stands for the 1st interaction of the 1st interaction flow diagram of the *Automobile* system. Interaction$_{11}$ is a type_1 interaction which describes the *Driver* actor interacts with the *Gas_Pedal* component.

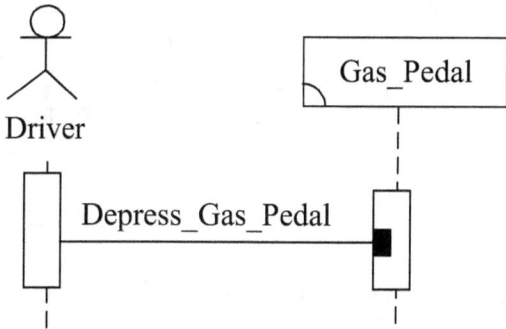

Interaction$_{12}$ stands for the 2nd interaction of the 1st interaction flow diagram of the *Automobile* system. Interaction$_{12}$ is a type_2 interaction which describes the *Gas_Pedal* component interacts with the *Engine* component.

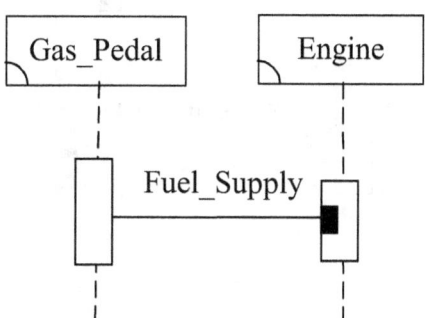

Interaction$_{21}$ stands for the 1st interaction of the 2nd interaction flow diagram of the *Automobile* system. Interaction$_{21}$ is a type_1 interaction which describes the *Driver* actor interacts with the *Brake_Pedal* component.

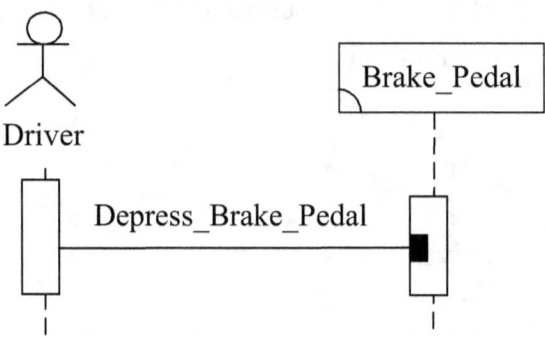

Interaction$_{22}$ stands for the 2nd interaction of the 2nd interaction flow diagram of the *Automobile* system. Interaction$_{22}$ is a type_2 interaction which describes the *Brake_Pedal* component interacts with the *Brake_Pad* component.

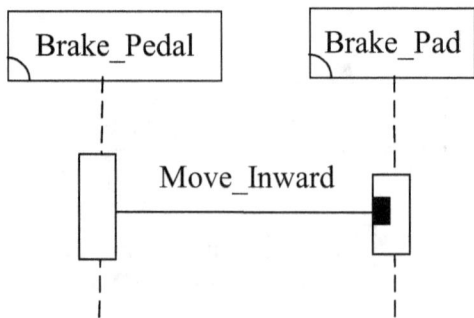

Interaction Flow Diagrams of the Automobile System

IFD_1 describes the 1st interaction flow diagram, i.e. *Accelerate_the_Car* behavior, of the *Automobile* system. IFD_1 is syntactically represented as "$Interaction_{11} \bullet Interaction_{12} \bullet STOP$".

$IFD_1 =$

$Interaction_{11} \bullet Interaction_{12} \bullet STOP$

IFD_2 describes the 2nd interaction flow diagram, i.e. *Stop_the_Car* behavior, of the *Automobile* system. IFD_2 is syntactically represented as "$Interaction_{21} \bullet Interaction_{22} \bullet STOP$".

$IFD_2 =$

$Interaction_{21} \bullet Interaction_{22} \bullet STOP$

Replicated Interaction Flow Diagrams of the Automobile System

Each interaction flow diagram may replicate itself a countably infinite time. Interaction$_{xyz}$ stands for the yth replication of the zth interaction of the xth interaction flow diagram.

!IFD$_1$ describes the infinite replication of the 1st interaction flow diagram, i.e. *Accelerate_the_Car* behavior, of the *Automobile* system. !IFD$_1$ is syntactically represented as "(Interaction$_{111}$●Interaction$_{112}$●*STOP*) ||
(Interaction$_{121}$●Interaction$_{122}$●*STOP*) ||
(Interaction$_{131}$●Interaction$_{132}$●*STOP*)   ..... ||
(Interaction$_{1\infty1}$●Interaction$_{1\infty2}$ ● *STOP*)".

!IFD$_1$ =

(Interaction$_{111}$ ● Interaction$_{112}$ ●*STOP*)
||
(Interaction$_{121}$ ● Interaction$_{122}$ ●*STOP*)
||
(Interaction$_{131}$ ● Interaction$_{132}$● *STOP*)
....
||
(Interaction$_{1\infty1}$ ● Interaction$_{1\infty2}$ ●*STOP*)

!IFD$_2$ describes the infinite replication of the 2nd interaction flow diagram, i.e. *Stop_the_Car* behavior, of the *Automobile* system. !IFD$_2$ is syntactically represented as "(Interaction$_{211}$●Interaction$_{212}$●*STOP*) ||
(Interaction$_{221}$●Interaction$_{222}$●*STOP*) ||
(Interaction$_{231}$●Interaction$_{232}$●*STOP*) ……. ||
(Interaction$_{2\infty1}$●Interaction$_{2\infty2}$●*STOP*)".

!IFD$_2$ =

(Interaction$_{211}$ ● Interaction$_{212}$ ● *STOP*)
||
(Interaction$_{221}$ ● Interaction$_{222}$ ● *STOP*)
||
(Interaction$_{231}$ ● Interaction$_{232}$ ● *STOP*)
…….
||
(Interaction$_{2\infty1}$ ● Interaction$_{2\infty2}$ ● *STOP*)

Infinite-Queue SBC Process of the Automobile System

The *Automobile* system's infinite-queue SBC process is syntactically represented as $(!IFD_1) \parallel (!IFD_2)$ which equals to "(Interaction$_{111}$•Interaction$_{112}$•*STOP*) $\parallel$ (Interaction$_{121}$•Interaction$_{122}$•*STOP*) $\parallel$ (Interaction$_{131}$•Interaction$_{132}$•*STOP*) ……. $\parallel$ (Interaction$_{1\infty1}$•Interaction$_{1\infty2}$•*STOP*) $\parallel$ (Interaction$_{211}$•Interaction$_{212}$•*STOP*) $\parallel$ (Interaction$_{221}$•Interaction$_{222}$•*STOP*) $\parallel$ (Interaction$_{231}$•Interaction$_{232}$•*STOP*) ……. $\parallel$ (Interaction$_{2\infty1}$•Interaction$_{2\infty2}$•*STOP*)".

Automobile $\overset{\text{def}}{=\joinrel=}$

$(\text{Interaction}_{111} \bullet \text{Interaction}_{112} \bullet STOP)$
$\parallel$
$(\text{Interaction}_{121} \bullet \text{Interaction}_{122} \bullet STOP)$
$\parallel$
$(\text{Interaction}_{131} \bullet \text{Interaction}_{132} \bullet STOP)$ ....
$\parallel$
$(\text{Interaction}_{1\infty1} \bullet \text{Interaction}_{1\infty2} \bullet STOP)$
$\parallel$
$(\text{Interaction}_{211} \bullet \text{Interaction}_{212} \bullet STOP)$
$\parallel$
$(\text{Interaction}_{221} \bullet \text{Interaction}_{222} \bullet STOP)$
$\parallel$
$(\text{Interaction}_{231} \bullet \text{Interaction}_{232} \bullet STOP)$ ....
$\parallel$
$(\text{Interaction}_{2\infty1} \bullet \text{Interaction}_{2\infty2} \bullet STOP)$

www.ingramcontent.com/pod-product-compliance
Lightning Source LLC
Chambersburg PA
CBHW072307200526
45168CB00014B/880